T0132764

FLORA OF TROPICAL EAST AFRICA

RANUNCULACEAE

E. Milne-Redhead and W. B. Turrill

Herbs or woody climbers or subshrubs. Leaves frequently compound with sheathing bases and most often without stipules, spirally arranged or opposite. Inflorescence generally terminal and many flowered, more rarely with few flowers, or these solitary. Flowers regular or irregular. Sepals free, hypogynous, sometimes petaloid. Petals sometimes absent, if present free, hypogynous, some nectariferous. Stamens indefinite in number, hypogynous. Carpels indefinite in number, few, or solitary, free or more or less joined together (if not solitary), with one or several ovules. Fruits achenes, drupelets, follicles, or capsules. Seeds without arils, with endosperm.

Flowers regular :
  Sepals more or less petaloid ; no nectariferous petals :
    Fruits dry achenes :
      Woody climbers or prostrate subshrubs: leaves
        opposite, sepals valvate .   .   .   .   1. **Clematis**
      Herbs or erect subshrubs: sepals more or less
        imbricate :
        Leaves opposite .   .   .   .   .   2. **Clematopsis**
        Leaves spirally arranged :
          Involucre absent .   .   .   .   3. **Thalictrum**
          Involucre present .   .   .   .   4. **Anemone**
    Fruit of fleshy drupelets .   .   .   .   5. **Knowltonia**
  Sepals sepaloid ; petals with nectaries .   .   .   6. **Ranunculus**
Flowers irregular .   .   .   .   7. **Delphinium**

1. CLEMATIS

L., Gen. Pl., ed. 5, 242 (1754)

Mostly woody climbing or suffruticose trailing plants. Leaves opposite, generally compound, petiole and rhachis capable of twining round supports but not ending in a naked tendril. Flowers regular, in many- to few-flowered panicles or solitary. Sepals 4 (rarely more), valvate, more or less petaloid. Petals and nectaries absent but staminodes, which gradually pass into the stamens, sometimes present. Carpels indefinite in number, normally uniovulate by abortion. Achenes capitate, with a persistent usually elongated plumose or naked style.

Leaflets ovate to ovate-lanceolate, entire to regularly
  dentate without lobes except sometimes 1–2 lobes on
  leaflets of upper leaves :
  Flowers large ; sepals 3·5–3·8 cm. long, erect at
    anthesis with the tips reflexed ; achenes (including
    persistent style) 9–10 cm. long .   .   .   1. *C. grandiflora*

Flowers small to medium ; sepals 0·7–1·4 cm. long,
  spreading at anthesis with tips not reflexed ;
  achenes (including persistent style) 2·5–4·0 cm.
  long . . . .   2. *C. simensis*
Leaflets suborbicular, ovate to oblong, unequally
  irregularly and often coarsely dentate, generally
  some or all of the leaflets lobed :
Plants often prostrate ; leaflets oblong . . .   3. *C. welwitschii*
Plants woody climbers ; leaflets ovate, broadly ovate
  to suborbicular in general outline :
Indumentum on underside of leaves (at least the
    young ones) soft and golden brown, and often
    dense ; flowering pedicels slender, about 4·5
    cm. long ; achenes (including persistent style)
    6 cm. long . . . .   4. *C. dolichopoda*
Indumentum on underside of leaves very dense, soft,
    and usually white or grey, or foliage glabrescent ;
    flowering pedicels generally less than 4 cm. long:
  Flower buds and back of sepals with yellow
    indumentum . . . . . .   5. *C. commutata*
  Flower buds and back of sepals with white, cream,
    or grey indumentum :
    Mature leaflets often asymmetric, trilobed, with
      coarse sharp teeth, the middle lobe long and
      sharply acuminate ; achenes (including
      persistent style) up to 6 cm. long . .   6. *C. viridiflora*
    Mature leaflets usually symmetric, either not
      markedly trilobed or if trilobed, the teeth
      more or less crenate and the central lobe not
      long and sharply acuminate ; achenes
      (including persistent style) up to 4 cm. long   7. *C. hirsuta*

1. **C. grandiflora** *DC.*, Syst. 1 : 151 (1818) ; F.T.A. 1 : 7 (1868) ; F.W.T.A.
1 : 67 (1927) ; Exell & Mendonça in Consp. Fl. Angol. 1 : 1 (1937) ; F.C.B.
2 : 186 (1951). Type : Sierra Leone, *Afzelius* (UPS, holo., BM, iso. !)

A tall woody climbing plant, up to 13 m. or more in length ; younger
stems with scattered hairs, becoming glabrous with age, ribbed and furrowed.
Leaves pinnate with usually 5 leaflets often unequal in size and reduced
when associated with the inflorescences ; leaflets ovate to oblong-ovate,
acute to shortly acuminate, more or less cordate at base, generally not
lobed, crenate-dentate but not strongly, pubescent to glabrous. Inflorescences with 1–3 flowers ; pedicels 3–4 cm. long ; flower buds ovoid to
conical-ovoid. Flowers campanulate. Sepals with sharply recurved tips,
about 3·5 cm. long, greenish yellow.

UGANDA. Mengo District : Namanve Forest, Mar. 1932, *Eggeling* 212 ! ; Kome Islands,
  Thomas 3774 !
DISTR. U2, 4 ; Sierra Leone to Belgian Congo and Angola
HAB. Margins and openings in forest, 1200 m.

SYN. *C. chlorantha* Lindl., Bot. Reg. t. 1234 (1829)
  C. pseudograndiflora O. Ktze. in Verh. Bot. Ver. Brand. 26 : 128 (1885). Type :
    Angola, *Welwitsch* 1218 (BM, holo. ! K, iso. !)

2. **C. simensis** *Fresen.* in Mus. Senckenb. 2 : 267 (1837) ; F.T.A. 1 :
6 (1868) ; F.C.B. 2 : 191 (1951). Type : Abyssinia, Semen, *Rüppell* (FR,
holo. !)

A tall climbing shrubby plant, up to 20 m. or more, often behaving as a strong liane ; younger stems more or less hairy but usually becoming glabrous or nearly so, longitudinally ribbed and furrowed. Leaves unequally pinnate with five leaflets, but frequently reduced in association with the inflorescence ; leaflets ovate to ovate-lanceolate, acuminate or shortly acuminate, rounded to cordate at base, entire to regularly dentate without lobes except sometimes 1–2 lobes on leaflets of uppermost leaves, nearly glabrous to densely pubescent on lower surface, a few scattered hairs on upper surface. Inflorescence many flowered ; pedicels, 1–3·5 cm. long ; flower buds ellipsoid. Sepals 7–16 mm. long, cream or white.

UGANDA. Kigezi District : Rubanda, June 1946, *Purseglove* 2064 !
KENYA. Chyulu Hills, 21 Apr. 1938, *Bally in C.M.* 7806 !
TANGANYIKA. Iringa District : E. Mufindi, 3 Aug. 1933, *Greenway* 3460 !
DISTR. U2–4 ; K3–5 ; T2, 3, 7 ; Abyssinia, A.-E. Sudan, Eritrea, Congo, Angola, Southern Rhodesia, Nyasaland, British Cameroons and Fernando Po
HAB. Forest edges and bushland, 1000—3360 m.

SYN. *C. orientalis* L. var. *simensis* (Fresen.) O. Ktze. in Verh. Bot. Ver. Brand. 26 : 126 (1885)
   C. sigensis Engl. in E.J. 45 : 271 (1910). Type : Tanganyika, E. Usambaras, *Warnecke* 452 (B, holo.)
   C. kissenyensis Engl. in Z.A.E. 207 (1911). Type : Belgian Congo, Kissenje, *Mildbraed* 1343 (B, holo., K, photo. !)
   C. altissima Hutch. in K. B. 1923 : 180 (1923). Type : Fernando Po, *Mann* 576 (K, holo. !)

VARIATION. There is a considerable range of variation within this species, particularly in leaf-shape, leaf-margin, indumentum and flower-size. Engler described a var. *kilimandscharica* from Marangu, 2700 m., *Volkens* 115, with ovate scarcely acuminate cordate leaflets. From Elgon and the area to the south-east of this mountain, plants with small flowers (sepals 7 mm. long) seem to be common. Material from Ruwenzori is very densely hairy. Until detailed field studies have been carried out we are averse to naming varieties within this very variable species.

   C. stolzii Engl. in E.J. 45 : 272 (1910) (Type : Tanganyika, Rungwe District, Masukulu, *Stolz* 161 (B, holo., K, iso. !, BM, iso. !)) is probably a hybrid with *C. simensis* Fresen as one of the parents.

3. C. welwitschii [*Hiern ex*] *O. Ktze.* in Verh. Bot. Ver. Brand. 26 : 171 (1885) ; Cat. Welw. Afr. Pl. 1 : 3 (1896) ; Exell & Mendonça in Consp. Fl. Angol. 1 : 3 (1937) ; F.C.B. 2 : 188, t. XV (1951). Type : Angola, Pungo Andongo, *Welwitsch* 1217 (BM, holo. !, K, iso. !)

A woody trailing or climbing plant ; younger stems pilose, glabrescent with age, longitudinally ribbed and furrowed. Leaves most often bipinnate or (rarely) pinnate with lateral leaflets more or less deeply trilobed ; leaflets typically oblong with few rather irregular teeth, glabrous or glabrescent on upper surface, more or less pilose, especially on veins, on lower surface. Inflorescences loose, usually about 5–7 flowered ; pedicels 1·5–6·5 cm. long ; buds ellipsoid, rounded or acute. Sepals 1·3–2·5 cm. long, outer surface pale pink with scattered hairs except at margin or glabrescent, inner surface densely lanate. Fig. 1, (p. 4).

TANGANYIKA. Mbeya District : Mbozi, 30 Aug. 1933, *Greenway* 3646 !
DISTR. T5, 7 ; Angola, southern Belgian Congo, Northern and Southern Rhodesia
HAB. Upland grassland, areas of abandoned cultivation, and *Brachystegia* woodland, 1500–1800 m.

SYN. *C. thunbergii* Steud. var. *angustisecta* Engl. in E. J. 30: 309 (1901). Type : Tanganyika, Mbeya District, Usafwa, Utengule, *Goetze* 1033 (B, holo., K, iso. !)
   C. kassneri Engl. in E. J. 45 : 274 (1910). Type : Belgian Congo, Katanga, *Kassner* 2663 (B, holo., K, iso. !)

FIG. 1. *CLEMATIS WELWITSCHII*—1, leaf, × 1; 2, inflorescence, × ½; 3, flower, × 1; 4, sepal, from within, × 2; 5, stamen, × 4; 6, carpel and style, × 4; 7, fruit × 1; 8, achene, × 4.

4. C. dolichopoda *Brenan* in K. B. 1949 : 71 (1949) & T.T.C.L. 464 (1949). Type : Tanganyika, Lushoto District, E. Usambaras, Nderema, *Scheffler* 82 (B, lecto.)

A climbing or scrambling plant ; younger stems covered with pale rusty or golden hairs, stems ribbed and furrowed. Leaves pinnate with usually 5 leaflets, much reduced in association with the inflorescences ; leaflets ovate to broadly or elongate ovate, not lobed or slightly trilobed, acute to slightly acuminate, base cordate or rounded, dentate-crenate, covered, often densely, with pale rusty or golden hairs on lower surface at least in younger leaves, with scattered hairs or glabrescent on upper surface. Inflorescences with 3–7 flowers ; pedicels 3·5–5 cm. long ; flower buds broadly ellipsoid, with yellow indumentum. Sepals 1·5–2·5 cm. long, cream.

TANGANYIKA. Lushoto District : E. Usambaras, Amani, 27 June 1929, *Greenway* 1630 ! ; Amani, *Warnecke* 422 ! ; Mt. Bomole, 30 June 1950, *Verdcourt* 275.
DISTR. **T3** ; endemic in Usambara Mts.
HAB. Edges of and clearings in lowland rain forest, 500–1000 m.

SYN. *C. longipes* Engl. in E. J. 45 : 273, fig. 2 C-E (1910), non Freyn (1880), *nom. illegit.*

5. C. commutata *O. Ktze.* in Verh. Bot. Ver. Brand. 26 : 128 (1885) ; Exell & Mendonça in Consp. Fl. Angol. 1 : 2 (1937) ; F.C.B. 2, 187 (1951). Type : Angola, Huila, *Welwitsch* 1215 (BM, holo. !, K, iso. !)

A climbing woody plant ; the younger stems pilose becoming more or less glabrous with age, longitudinally ridged and furrowed. Leaves pinnate, with usually 5 leaflets ; leaflets ovate to lanceolate-ovate, often strongly asymmetric, curved in upper part (especially the terminal leaflet), with or without slight lobes, irregularly and unequally dentate to almost entire, glabrous on upper surface, somewhat shortly pilose on lower surface, particularly on the veins. Inflorescences with 1–3 flowers ; pedicels 1–2 cm. long ; buds broadly ovoid to spherical, with yellow indumentum. Sepals 1·7–2·5 cm. long, cream, more or less pilose to glabrescent on inner surface.

UGANDA. Kigezi District : Luhiza, Kinaba, Mar. 1947, *Purseglove* 2349 !
TANGANYIKA. Iringa District : Iringa, *Emson* 517 !
DISTR. **U2** ; **T4, 7** ; Belgian Congo, Ruanda-Urundi and Angola
HAB. Wooded grassland and upland evergreen bushland, 1600–1860 m.

SYN. *C. iringaënsis* Engl. in E. J. 28 : 388 (1900). Type : Tanganyika, Iringa, *Goetze* 705 (B, holo., K, iso. !)
C. keilii Engl. in E. J. 45 : 273 (1910). Type : Ruanda, Kagera, *Keil* 110 (B, holo.)

6. C. viridiflora *Bertol.*, Miscell. Bot. 19 : 7, t. 3 (1858). Type : Portuguese East Africa, Inhambane, *Fornasini*, (BO, holo. †)

A tall climbing shrubby plant ; younger stems with short curved hairs but becoming glabrous or nearly so, longitudinally ribbed and furrowed. Leaves pinnate, with usually 5 leaflets, reduced in association with the inflorescences ; leaflets ovate to broadly ovate in general outline, obliquely asymmetric with acuminate lobes with relatively few and large asymmetric (curved) acute teeth, slightly cordate to broadly cuneate, with scattered hairs on lower surface, upper surface nearly glabrous, well formed leaflets 7–12 cm. long and 5–10 cm. broad. Inflorescences many flowered ; pedicels 1–5 cm. long ; flower buds broadly ovoid. Sepals 1·2–1·8 cm. long, cream.

ZANZIBAR. Chwaka, 25 Aug. 1926, *Toms* 42 ! ; Makunduchi Road, 19th mile, 7 July 1950, *Williams* 40 ! ; Chwaka Road, mile 17–18, 13 Aug. 1950, *Williams* 62 !
DISTR. **Z** ; Portuguese East Africa
HAB. Climbing up trees in coastal evergreen bushland, below 250 m.

SYN. [*C. zanzibarensis* Boj. ex Loud., Hort. Brit. 228 (1830) ; Oliv., F.T.A. 1 : 8 (1868) ;
  Williams, U. O. P. Z. 198 (1949), *nomen nudum*]
  [*C. zanzibarica* Sweet, Hort. Brit. ed. 2, 1 (1830), *nomen nudum*]

It is possible that several specimens from the coastal region of Tanganyika should
be placed in this species. We have, however, not seen material sufficiently well collected
for definite determination.

7. C. hirsuta *Perr. & Guill.* in Fl. Seneg. Tent. 1 : 1 (1831) ; Exell & Mendonça in Consp. Fl. Angol. 1 : 3 (1937) ; F.C.B. 2, 192 (1951). Type : Cape Verde Peninsula, *Leprieur* (P, holo.)

A tall climbing shrubby plant (more rarely decumbent in grassland),
1–4 m. or more tall ; younger stems more or less softly hairy but usually
becoming glabrous or nearly so, longitudinally ribbed and furrowed. Leaves
pinnate with 5 leaflets, rarely bipinnate, frequently well developed leaves are
associated with the inflorescence ; leaflets suborbicular to ovate in general
outline, acuminate, shortly acuminate, acute, or subobtuse, cordate to
rounded rarely truncate at base, most often a longer central lobe with a
shorter lateral one on each side, crenate-dentate with teeth of different
sizes, from almost glabrous to lanate on lower surface, glabrous to more or
less pubescent on upper surface. Inflorescences generally many-flowered ;
pedicels 1–3 cm. long ; flower buds spherical to ellipsoid, rounded to acuminate. Sepals 1–2·7 cm. long, cream or white.

UGANDA. Mengo District : mile 8 Masaka Road, Sept. 1937, *Chandler* 1930 !
KENYA. Kericho District : Sotik Hills, 22 Aug. 1946, *Greenway* 7851 !
TANGANYIKA. Mbulu District : Mbulumbulu, 13 July 1943, *Greenway* 6764 !
DISTR. U1–4 ; K3–7 ; T2, 4, 5, 7, 8 ; widely spread in most parts of Tropical Africa
HAB. Forest edges and openings and in wooded grassland, 1000–3000 m.

SYN. *C. glaucescens* Fresen. in Mus. Senckenb. 2 : 268 (1837). Type : Abyssinia,
  Rüppell (FR, holo. !)
  C. inciso-dentata A. Rich., Tent. Fl. Abyss. 1 : 2 (1847). Type : Abyssinia,
  Choa, *Petit* (P, holo. !)
  C. petersiana Klotzsch in Peters, Mossamb. Bot. 1 : 170 (1861). Type ; Portu-
  guese East Africa, Tette, *Peters* (B, holo.)
  [*C. grata* Wall. sec. Oliv., F.T.A. 1 : 7 (1868), non Wall.]
  C. friesiorum Ulbr. in N. B. G. B. 10 : 914 (1930). Type : Kenya : Mt. Kenya
  between Jaracuma and Meru, *Fries* 1598 (UPS, holo., K, iso. !)

VARIATION. As accepted here this is an extremely variable species, especially in leaflet
  shape and size, leaf-margin, indumentum and flower size. The commonest East
  African variant has leaflets averaging 3 cm. long and 2·5 cm. broad, with a dense
  indumentum on the lower surface. Engler described a var. *pilosissima*, in E. J. 30 :
  309 (1902) (*Goetze* 1283, E.A., isotype !) under the specific name *C. wightiana* Wall.
  with all the parts shortly and densely grey-pilose. An extreme variant is *Purseglove*
  3372 from Uganda, Kigezi District, Bukimbiri, 1950 m., May 1950. This has leaflets
  with blades up to 9 cm. long and 6·5 cm. broad, with dense golden yellow indumentum
  on the under surface. Owing to the reticulation of characters it has not been found
  possible to divide the species, as here accepted, into clear-cut varieties. It may be
  suggested that what were at one time two or more distinct species have amalgamated
  by interbreeding and back-crossing in Tropical Africa to produce the variable populations assigned here to *C. hirsuta*, and which may, in origin, represent a hybrid swarm.
  The attention of botanists is called to the need for intensive field studies on populations of this variable species.

C. hirsuta is very closely related to the Indian *C. wightiana* [Wall. ex] Wight,
Prod. Fl. Penins. Ind. Or. 2 (1834), but this latter has a yellow, not cream or white,
indumentum to the sepals (as well seen in bud).

2. CLEMATOPSIS

[Boj. ex] Hutch. in K.B. 1920 : 12–22 (1920)

Perennial herbs with erect stems which are rarely somewhat woody in
lower part. Leaves opposite, simple or more often compound, petiole and

rhachis not acting as tendrils. Flowers regular, solitary or few at the ends of stems or branches. Sepals 4 (rarely more), more or less imbricate, petaloid. Petals and nectaries absent. Carpels indefinite in number, with one fertile ovule. Achenes capitate, with a persistent elongated plumose style.

Leaves all simple .   .   .   .   1. *C. uhehensis*
Leaves compound (sometimes some simple leaves near
   the base or apex of stems) .   .   .   .   .   2. *C. scabiosifolia*

1. **C. uhehensis** (*Engl.*) *Staner & Léonard* in B.S.B.B. 82 : 342 (1950). Type : Tanganyika, Iringa District, Utschungwe Mts. near Kissinga, *Goetze* 579 (B, holo. !)

Perennial, with erect stems sometimes woody at the base, 4–9 dm. tall, strongly longitudinally striate, with spreading shaggy hairs. Leaves simple, ovate, up to 9 cm. long and 5·5 cm. broad, sessile or shortly petioled, coarsely and somewhat irregularly dentate, or sometimes obscurely lobed, with distinct scattered hairs. Flowers mostly solitary at the ends of the stems, 6·5–12 cm. in diameter. Sepals spreading, softly hairy on both surfaces, white tinged mauve to purplish rose. Fruit unknown.

TANGANYIKA. Rungwe District : Mt. Rungwe, 11 Mar. 1932, *St. Clair-Thompson* 831 !
DISTR. **T7** ; endemic in Southern Highlands of Tanganyika
HAB. Upland and wooded grasslands, 1800–2800 m.

SYN. *Clematis uhehensis* Engl. in E. J. 28 : 387 (1900)
    Clematopsis simplicifolia Hutch. & Summerh. in K.B. 1925 : 361 (1925). Type :
    Tanganyika, Rungwe Mt., *Stolz* 2514 (K, holo. !)

2. **C. scabiosifolia** (*DC.*) *Hutch.* in K.B. 1920 : 20 (1920) ; Exell & Mendonça, Consp. Fl. Angol. 1 : 5 (1937) ; F.C.B. 2 ; 198, t. XVI (1951). Type : probably from Angola, collector uncertain. (P, holo.)

Perennial, with erect herbaceous stems 0·7–1·5 m. tall, strongly longitudinally striate, indumentum varying from densely silky matted hairs to spreading not silky hairs, or sparsely spreading hairs, or glabrescent. Leaves pinnate to bipinnate or trifoliolate, except sometimes near the base or apex of the stems where they may be simple, very variable in details of shape and indumentum (see key to groups below). Flowers solitary to fairly numerous at the ends of stems or main branches, 3·5–7 cm. in diameter. Sepals more or less spreading, softly hairy on both surfaces, white, cream, mauve, or pink. Anthers up to 5 mm. long. Achenes in heads up to 10 cm. in diameter.

Seven groups are recognized by Exell, Léonard, and Milne-Redhead, in B.S.B.B. 83 : 412 (1951), as occurring in Africa. The following key is to those groups so far recorded from Tropical East Africa :

KEY TO THE GROUPS OF *CLEMATOPSIS SCABIOSIFOLIA*

Stems 1–3-flowered ; flowers usually solitary ; internodes
   usually exceeding the leaves .   .   .   .   .   .   Group D
Stems usually many-flowered ; at least the lower leaves longer
   than the internodes :
Sepals markedly acuminate .   .   .   .   .   .   Group G
Sepals not markedly acuminate :
   Leaves densely sericeous-tomentose, the indumentum
     usually concealing the tertiary nervation .   .   .   Group F
   Leaves pubescent or sparsely pubescent, pubescence mainly
     on the nerves, tertiary nervation visible :

Pinnae suborbicular in outline, crenate-serrate, usually
rounded at the base .   .   .   .   .   .   Group C
Pinnae ovate to elliptic, irregularly serrate or incised,
tending to be cuneate at the base .   .   .   .   Group B

Group B

TANGANYIKA.   Kilosa District : Usagara Mts., Uponera, *Busse* 295 !
DISTR.   T6 ; Northern Rhodesia, Belgian Congo

SYN.   *Clematis busseana* Engl. in E. J. 45 : 269 (1910).   Type : *Busse* 295 (B, holo. !
     K, iso. !)

Group C

TANGANYIKA.   Njombe District : Lupembe, watershed of upper Ruhudji, Jan. 1931,
     Schlieben 64 !
DISTR.   T7 ; Nyasaland, Portuguese East Africa

SYN.   *Clematis kirkii* Oliv., F.T.A. 1 : 5 (1868).   Type : Nyasaland, Manganja Hills,
     Kirk (K, holo !)
     Clematopsis kirkii (Oliv.) Hutch. in K.B. 1920 : 17 (1920)

Group D

UGANDA.   West Nile District : Lindu, 18 Mar. 1945, *Greenway & Eggeling* 7228 !
KENYA.   West Suk District : Kapenguria, Mar. 1935, *Thorold* 3211 !
TANGANYIKA.   Bukoba District : Ndama and Mabira, Oct. 1931, *Haarer* 2251 !
DISTR.   U1–3 ; K2, 3, 5 ; T1, 7 ; Nigeria, British Cameroons, French Cameroons,
     A.-E. Sudan, Nyasaland, Northern Rhodesia

SYN.   *Clematopsis oliveri* Hutch. in K.B. 1920 : 20 (1920).   Type : A.-E. Sudan,
     White Nile, *Petherick* (K, holo. !)
     An extreme variant of this group, with the leaves and leaf segments long and narrow,
is represented by *Clematopsis lineariloba* Hutch. & Summerhayes, in K.B. 1925 : 361,
(1925) based on *Stolz* 2385 (K, holo. !), from Mbeya District, Usafwa, Dec. 1913.

Group F

UGANDA.   Kigezi District : May 1933, *Mrs. Sandford in Tothill* 1212 !
TANGANYIKA.   Bukoba District : south of Nyakahanga, Oct. 1931, *Haarer* 2252 !
DISTR.   U2, T1, 2, 5, 7 ; Angola, Northern Rhodesia, Belgian Congo

SYN.   *Clematis stuhlmannii* Hieron. in P.O.A. C : 180 (1895).   Type : Tanganyika,
     Bukoba District, Kagehi, *Stuhlmann* 3491 (B, holo. !, K, iso. !)
     Clematopsis stuhlmannii (Hieron.) Hutch. in K.B. 1920 : 20 (1920)

Group G

TANGANYIKA.   Njombe District : Msima Stock Farm, 1932, *Emson* 375 !
DISTR.   T7 ; Nyasaland, Belgian Congo

Specimens showing characters or combinations of characters intermediate between
those of two or more groups are of fairly frequent occurrence and to some of these names
have been given.   An example is *Clematis goetzei* Engl. in E.J. 28 : 388 (1900) (Type :
Tanganyika, Iringa District, Utschungwe [Mts.], Mufindi-Dabaga, Feb. 1899, *Goetze* 639
(B, holo. !)), which is intermediate between groups F and G.   Groups A and E are not
recorded from Tropical East Africa, but some specimens with some of the characters of
A have been collected.

HAB. (for aggregate species).   Upland grassland, *Brachystegia* woodland and abandoned
cultivated land, 1080 to 2000 m.

SYN. (for aggregate species).   *Clematis scabiosaefolia* DC., Syst. 1 : 154 (1818)

3. THALICTRUM

L., Gen. Pl., ed. 5, 242 (1754)

Herbs with compound, spirally arranged leaves, with sheathing bases,
often with stipules and stipels.   Flowers relatively small, generally in

paniculate inflorescences. Sepals 3–5, imbricate in bud, green or petaloid, caducous. Petals and nectaries absent. Stamens 3 to indefinite in number. Carpels 1 to indefinite in number, uniovulate. Achenes stipitate or sessile, with a persistent or deciduous style.

It is possible that all the species (if such they be) recorded below, except *T. zernyi*, belong to one section of the genus, characterized by very elongated pedicels in the infructescences. Unfortunately, fruiting material is not available for all the species and fruiting and flowering material from the same plant is practically non-existent in herbaria. The attention of field workers is called to the need for such material.

It should be noted that *T. minus* L. in the broad sense has been recorded from Abyssinia and from South Africa and may occur in mountainous areas in East Africa.

Carpels 10–14 . . . . 1. *T. zernyi*
Carpels 1–4 :
  Inflorescence compact ; pedicels 2–5 mm. long . 2. *T. stolzii*
  Inflorescence lax ; pedicels 5–50 mm. long :
    Sepals 3–4·5 mm. ; stamens 8–23 ; prolongation
      of anther connective 0·6–0·8 mm. . . 3. *T. boivinianum*
    Sepals 1·5–4 mm. ; stamens 3–13 ; prolongation
      of anther connective 0·05–0·4 mm. . . 4. *T. rhynchocarpum*

1. **T. zernyi** *Ulbr.* in N. B. G. B. 15 : 715 (1942). Type : Tanganyika, Matengo highlands, *Zerny* 268 (W, holo. !)

Perennial herb with erect stems 1·5–5 dm. tall, glabrous, simple or branched low down. Leaves short, 1·5–6 cm. in total length (including petiole), bipinnate ; leaflets elliptic to oblate, often trilobed, apices of lobes more or less rounded and very slightly apiculate, bases slightly cuneate to truncate. Inflorescences few-flowered, in young state compact, pedicels in infruct-escence scarcely elongated and filiform. Sepals 8 mm. long, white on inside, rose-coloured on outside. Stamens about 15 ; anther thecae 2·3 mm. long, prolongation of connective up to 0·2 mm. long. Carpels 10–14 ; stigma 1·5 mm. long at anthesis, curved or curled. Achenes sessile, without accrescent stipes.

TANGANYIKA. Rungwe District : Ukinga Mts., Mwakalele, 8 Jan. 1914, *Stolz* 2421 !
DISTR. **T4, 7, 8** ; endemic in southern Tanganyika
HAB. Rocky places in wooded grassland, 1800—2200 m.

2. **T. stolzii** *Ulbr.* in N. B. G. B. 10 : 916 (1930). Type : Tanganyika, Njombe District, Bulongwa, *Stolz* 2175 (B, holo. †, K, iso. !)

Perennial herb with erect stems 1·25–1·5 m. tall, glabrous, somewhat branched. Leaves moderately large, middle ones about 2 dm. long, bi-pinnate to quadripinnate ; leaflets broadly ovate to oblate in general outline sometimes trilobed or with few coarse teeth, rounded to subacute and shortly apiculate, base rounded or cordate. Inflorescences rather small, compact, and apparently few flowered ; pedicels 2 to 5 mm. long. Sepals 4 mm. long. Stamens 5–10 ; anther thecae 1·2 mm. long, prolongation of connective 0·1 mm. long. Carpels 1–2 ; stigma 3–4 mm. long at anthesis. Fruit unknown.

TANGANYIKA. Njombe District : Bulongwa, 17 Sep. 1913, *Stolz* 2175 !
DISTR. **T7** ; endemic in southern Tanganyika
HAB. Moist bamboo thicket about 2100 m.

Fig. 2. *THALICTRUM BOIVINIANUM*—1, leaf, × ½; 2, base of flowering branch, showing glands, × 10; 3, inflorescence, × 1; 4, flower, × 6; 5, pistil, × 6.

3. **T. boivinianum** *Staner & Léonard* in B.J.B.B. 19 : 450 (1949) ; F.C.B. 2 : 183 (1951). Type : Belgian Congo, Kivu District, mountains S.W. of Lemera, *Chapin* 518 (BR, holo. !).

Perennial herb, with erect stems 0·8 to 2 m. tall and much branched in the inflorescence region. Leaves large, up to 3 dm. or more in length (including petiole), finely divided, quadripinnate ; leaflets in outline elliptic to oblate, most often with 3 lobes which may be dentate, sometimes more or less nearly entire, apices of lobes rounded to acute, shortly apiculate, bases rounded to slightly cuneate or slightly cordate, sometimes asymmetric. Inflorescence very lax, often large and much branched. Sepals 3·0–4·5 mm. long, purple to mauve. Stamens 8–23 ; anther thecae 2·2–2·7 mm. long, prolongation of connective 0·6–0·8 mm. long. Carpels 1 to 4 ; stigma 6 to 7 mm. long at anthesis. Fruit unknown. Fig. 2.

UGANDA. Ruwenzori, Bujuku Valley, Nyamuleju, Aug. 1933, *Eggeling* 1269 !
KENYA. W. slopes of Mt. Kenya, *Mearns* 1350
DISTR. U2 ; K4 ; Belgian Congo
HAB. Moist bamboo thicket and upland moor, 2700–3630 m.

4. **T. rhynchocarpum** *Dillon & A. Rich.* in Ann. Sci. Nat., sér. 2, 14 : 262 (1840) ; F.T.A. 1 : 8 (1868) ; Weimarck in Svensk Bot. Tidskr. 30 : 46 (1936) ; F.C.B. 2 : 181 (1951). Type : Abyssinia, Tigré, near Adowa, *Quartin-Dillon* (P, holo. !)

Perennial herb with erect stems 1–4 m. tall, more or less branched. Leaves rather large, up to 4 dm. or more in length (including petiole), tripinnate to quadripinnate ; leaflets elliptic-oblong to broadly ovate and oblate, sometimes rather obscurely trilobed, with coarse teeth, apices rounded, usually apiculate, bases rounded to cordate. Inflorescence lax, many flowered. Sepals 1·5–4·0 mm. long, green to purplish. Stamens 3–13 ; anther thecae 0·8–1·7 mm. long, prolongation of connective 0·05–0·4 mm. long. Carpels 1–2 ; stigma 2–7 mm. long at anthesis. Achenes on slender elongated pedicels which are 6–11 cm. long, asymmetric, tapering into a persistent style, at the base tapering to a stipe or nearly sessile.

UGANDA. Kigezi District : Kachwekano Farm, May 1949, *Purseglove* 2814 !
KENYA. Elgon, May 1931, *Lugard* 660 !
TANGANYIKA. Kilimanjaro, S. slope between the Umbwe and Weru Weru rivers, 14 Aug. 1932, *Greenway* 3025 !
DISTR. U1–3 ; K3–5 ; T2, 3, 4, 6, 7 ; widely distributed in tropical Africa from Abyssinia to Cameroons and Fernando Po and south to Natal
HAB. Glades and undergrowth of upland rain forest, upland evergreen bushland on stream sides and upland grassland, 1620–3150 m.

SYN. *T. mannii* Hutch. in K.B. 1927 : 154 (1927) and F.W.T.A. 1, 66 (1927). Type : Fernando Po, *Mann* 293 (K, holo. !)
    T. chapinii B. Boiv. in Rhodora 46 : 395 (1944), *pro parte*. Type : Belgian Congo, Kivu District, Mt. Karisimbi, *Chapin* 386 (NY, holo.)
    T. impexum B. Boiv. l.c. 395 (1944). Type : Kenya, between Naiok River and Lake Naivasha, *Mearns* 630 (US, holo.)
    T. innitens B. Boiv. l.c. 394 (1944). Type : Cape Province, " Kabousie," *Murray* 598 (GH, holo.)

4. ANEMONE

L., Gen. Pl., ed. 5, 241 (1754)

Perennial herbs with lobed or dissected basal leaves. Often with an involucre of a whorl of 3 leaves (often very much reduced) below the flower. Flowers regular. Sepals petaloid, variable in number. Petals absent. Stamens indefinite in number, sometimes the outer ones staminodal. Carpels indefinite in number, uniovulate. Ovule pendulous. Achenes with persistent, naked to plumose, styles.

Anemone thomsonii *Oliv.* in J. L. S. 21 : 397 (1885) ; Hook., Ic. Pl. t. 1491 (1885). Type : Tanganyika, Kilimanjaro, *J. Thomson* (K, holo. !)

Rhizome short and stout. Basal leaves usually thrice ternate, the ultimate lobes more or less deeply and narrowly segmented, 0·3–4·0 dm. long, more or less hairy on lower surface. Flowering stems 0·7–7·0 dm. tall, with a silky adpressed indumentum dense near the involucre and below the flower, one-flowered. Involucre very reduced, sometimes only slightly foliaceous. Flower 2·5–6·5 cm. in diameter. Sepals white or white tinged pink on inner surface, outer surface often pink, red, or purple. Carpels and achenes with dense silky indumentum, style short and glabrous, less than 1 mm. long in fruit.

var. thomsonii

Basal leaves with leaf segments cuneately broadly elliptic or oblate, more or less deeply lobed or segmented with the ultimate lobes or segments 1·5–5 mm. broad.

UGANDA. Mbale District : Elgon, above and below Mudangi, *Liebenberg* 1605 !
KENYA. Elgon, July 1933, *Dale* 3090 !
TANGANYIKA. Kilimanjaro, 5 Sept. 1929, *Cotton* 16 ! ; Peter's Hut to Bismarck Hill, 25 Febr. 1934, *Greenway* 3777 !
DISTR. U3 ; K4, 5 ; T2 ; Abyssinia, A.-E. Sudan, Belgian Congo

var. friesiorum *Ulbr.* in N. B. G. B. 10 : 910 (1930). Type : Aberdare Mts., near Sattimma, *Fries* 2366 (UPS, holo.)

Segments of basal leaves cuneately obliquely ovate-elliptic or oblong elliptic, more or less deeply lobed, ultimate lobes (not teeth) 4–8 mm. broad.

KENYA. Aberdare Mts., *Battiscombe* 533.
DISTR. K3, 4 ; endemic in Aberdare Mts.

var. angustisecta *Milne-Redhead & Turrill* in K. B. 1950 : 389 (1951). Type : Tanganyika, Mt. Hanang, *Greenway* 7639 (K, holo. !)

Leaves extremely dissected with ultimate segments linear and 0·5–0·75 mm. broad.

TANGANYIKA. Mbulu District : Mt. Hanang, 2 Sept. 1932, *B. D. Burtt* 4011 !
DISTR. T2 ; endemic on Mt. Hanang

HAB. (for species as whole). Moist rocky ledges and crevices, upland moor and moor grassland, 2700 to 3900 m.
DISTR. (of species as whole). U3 ; K3–5 ; T2 ; Abyssinia, southern A.-E. Sudan (Imatong Mts.) and Belgian Congo.

5. KNOWLTONIA

Salisb., Prodr. 372 (1796)

Perennial herbs with compound leaves and multi-flowered inflorescences. An involucre or reduced leaves at the inflorescence branches. Flowers regular. Tepals more or less petaloid, about 15 but variable in number, the outer often somewhat smaller and slightly sepaloid, none with nectariferous pits. Stamens indefinite in number. Carpels indefinite in number, uniovulate, with glabrous deciduous styles. Fruits composed of fleshy drupelets.

K. transvaalensis *Szyszyl.*, Polypet. Thalam. Rehm. 99 (1887). Type : Transvaal, Houtbosh, *Rehmann* 6402 (?, holo., K, iso. !)

Short stout rhizome. Basal leaves biternate to ternate, the segments varying in size and lobing, lobes 2–10 cm. long, 0·8–5 cm. broad ; petioles 7–17 cm. long, densely lanate-pubescent towards the base, especially on the sheathing part, with spreading hairs, or rarely glabrous above. Flowering stems 3–8 dm. tall, with dense lanate pubescence towards the base, usually shortly densely pubescent in inflorescence parts. Flowers simply or compoundly " umbellate," 2–15, usually 2·5–3 cm. in diameter, sweet-scented. Tepals white, sometimes outer tinged with purple. Carpels glabrous.

TANGANYIKA. Njombe District : Lupembe, 1 Oct. 1931 *Schlieben* 1242 !
DISTR. **T7** ; Nyasaland, Southern Rhodesia, Transvaal.
HAB. Upland grassland in damp places ; 1600 m.

SYN. *Anemone whyteana* Bak. f. in Trans. Linn. Soc. ser. 2, 4, 4 (1894). Type : Nyasa-
    land, Mt. Mlanje, *Whyte* 100 (BM, holo. ! K, iso. !)
   A. peneënsis Bak. f. in J. L. S. 40 : 16 (1911). Type : Southern Rhodesia,
    Mt. Pene, *Swynnerton* 783 (BM, holo. ! K, iso. !)
   A. transvaalensis (Szyszyl.) Burtt-Davy in Ann. Transv. Mus. 3 : 121 (1912)
   Knowltonia whytei Engl. in V.E. 3 (1) : 170 (1915), *in obs.* ; *sphalm. pro K.
   whyteana* (Bak. f.) Engl.
   K. multiflora Burtt-Davy in K. B. 1921 : 343 (1921) ; Burtt-Davy, Man. Fl. Pl.
    Transv. 1 : 110 (1926). Type : Transvaal, Lydenburg District, Mac-a-Mac,
    Mudd (K, holo. !)

6. RANUNCULUS

L., Gen. Pl., ed. 5, 243 (1754)

Herbs with simple or compound spirally arranged leaves, usually ex-stipulate. Flowers regular, in one- to many-flowered inflorescences, without definite involucres of bracts or reduced leaves. Sepals spreading or reflexed in anthesis. Petals 5–8 or more, white, yellow, or with anthocyanin colours, with basal nectariferous pit with or without a scale. Stamens indefinite in number. Carpels indefinite in number, uniovulate. Achenes with persistent, glabrous, sometimes hooked style.

Leaves entire to dentate, not lobed . . . . 1. *R. volkensii*
Leaves compound or deeply cut or lobed :
  Leaves simply but definitely multipinnate . . 2. *R. oreophytus*
  Leaves not simply multipinnate :
    Sepals spreading in anthesis :
      Leaves pinnatilobed or pinnatisect ; petals white 3. *R. stagnalis*
      Leaves trisect, trilobed, trifoliolate, palmatisect,
        or palmatilobed :
        Roots thickened and more or less elongate-
          tuberous ; stems erect ; fruiting pedicels
          erect ; petals yellow . . . . 4. *R. raeae*
        Roots fibrous ; stems prostrate ; fruiting
          pedicels reflexed :
          Leaves trifoliolate ; terminal leaflet generally
            petiolate . . . . . . 5. *R. aberdaricus*
          Leaves trisect ; segments with cuneate, often
            rather broad, bases . . . . 6. *R. cryptanthus*
    Sepals reflexed in anthesis :
      Basal leaves trifoliolate :
        Leaflets of basal leaves with margin coarsely
          dentate and more or less deeply lobed ;
          terminal leaflet 0·9–2·5 cm. long, 1·0–2·9 cm.
          broad . . . . 7. *R. keniensis*

Leaflets of basal leaves with margins finely
serrate ;   terminal leaflet 7–8 cm. long,
6–6·5 cm. broad   .   .   .   .   . 8. *R. bequaertii*
Basal leaves bi- or tri-pinnatisect   .   .   . 9. *R. multifidus*

1. **R. volkensii** *Engl.* in P. O. A. C : 181 (1895) ; Ulbr. in N. B. G. B. 10 :
901 (1930) ; Staner in B. J. B. B. 15 : 310 (1939) ; Robyns, F.P.N.A. 1 :
171 (1948) ; F.C.B. 2 : 171 (1951). Type : Tanganyika, Kilimanjaro,
Marangu, *Volkens* 971 (B, holo. †, K, iso. !)

Perennial herb with usually elongated slender prostrate stems, rooting at
the nodes whence arise tufts of leaves. Leaves not lobed or otherwise
divided, lanceolate to broadly ovate or even reniform, apex rounded, obtuse,
acute, or acuminate, base cordate, truncate, to slightly and broadly cuneate,
margin dentate to nearly entire. Flowers single at the nodes, pedicels
1–8·5 cm. long, glabrous or adpressed hirsute. Sepals spreading, glabrous.
Petals 4 to 7 mm. long, yellow. Achenes more or less fusiform, smooth.

UGANDA. Kigezi District : Muhavura–Mgahenga saddle, Sept. 1946, *Purseglove* 2163 !
KENYA. Mt. Kenya (west), near Forest Station, 3 Jan. 1922, *Fries* 684 !
TANGANYIKA. Kilimanjaro, near Peter's Hut, 22 Feb. 1934, *Greenway* 3737 !
DISTR. **U**2, 3 ; **K**3, 4 ; **T**2 ; Ruanda-Urundi and Belgian Congo
HAB. Marshy places, edge of streams, and other wet places in upland moor and moor
grassland, 2700 to 4050 m.

SYN. *R. ulbrichii* Engl. in Z.A.E. 208 (1911) ; Staner in B. J. B. B. 15 : 310, tab. XI.
fig. 2 (1939) ; Robyns, F. P. N. A. 1 : 172 (1948) ; F. C. B. **2** : 172 (1951).
Type : Ruanda-Urundi, Rugege forest, *Mildbraed* 982a (B, holo. †)
R. elgonensis Ulbr. in N. B. G. B. 10 : 902 (1930). Type : Uganda, Elgon,
Dummer 3312 (B, holo. †, K, iso. !)
R. helogeton Ulbr. l.c. 901 (1930). Type : Kenya, Aberdare Mts., near Sattimma,
Fries 2352 (UPS, holo., K, iso. !)

2. **R. oreophytus** *Del.* in Ann. Sci. Nat. sér. 2, 20 : 89 (1843) ; F.T.A. 1 :
10 (1868) ; Ulbr. in N. B. G. B. 10 : 905–9 (1930) ; F.C.B. 2 : 173 (1951).
Type : Abyssinia, Semen, Silké, *Galinier* (P, holo.)

Perennial herb, usually acaulescent, and with much elongated somewhat
thickened roots. Leaves about 5–12, forming a rosette, pinnate with 2–4
pairs of leaflets and an odd terminal one ; leaflets more or less elliptic or
broadly elliptic in general outline, the lateral ones mostly with 1 or 2 coarse
teeth or lobes on each margin, the terminal with several lobes or teeth, apex
and teeth acute to subacute, pilose to nearly or quite glabrous ; rhachis and
petiole more or less pilose. Flowers arising singly from among the basal
leaves, rarely from elongated stems ; pedicels 0·5–10 cm. long, adpressed
pilose. Sepals spreading, glabrous to slightly hairy. Petals generally 5,
1·0–2·5 cm. long, 2·5–5 mm. broad, yellow. Achenes in spherical heads
which by curvature of the pedicels become buried in the soil (geocarpic),
with no thickened margin and smooth rounded sides ; persistent beak up to
1·5 mm. long, slender, more or less straight. Fig. 3.

UGANDA. Ruwenzori, Bujuku Valley, Aug. 1933, *Eggeling* 1300 !
KENYA. Mt. Kinangop, Loreko, 18 July 1931, *Napier* 1249 !
TANGANYIKA. Kilimanjaro, near Peter's Hut, 23 Feb. 1934, *Greenway* 3768 !
DISTR. **U**2, 3 ; **K**3, 4 ; **T**2, 3, 6, 7 ; Abyssinia, southern A.-E. Sudan, eastern Belgian
Congo
HAB. Rock clefts and short grass near streams, and other wet and boggy places, in
upland moor, 2240–4350 m.

SYN. [*R. tenuirostris* Steud. ex Hochst. in Flora 27 : 97 (1844), *nomen nudum*]
R. gunae Schweinf. in Verh. Zool.–bot. Ges. Wien 18 : 666 (1868). Type :
Abyssinia, Mt. Guna, *Steudner* 1207 (B, holo. †)

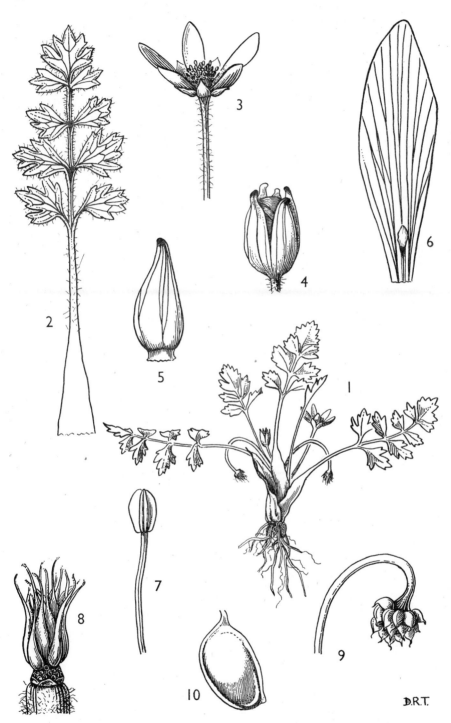

FIG. 3. *RANUNCULUS OREOPHYTUS*—1, whole plant, × ½; 2, leaf, × 1; 3, flower, × 1; 4, flower-bud, × 3; 5, sepal, × 4; 6, petal, × 4; 7, stamen, × 6; 8, pistil, × 6; 9, fruit, × 2; 10, achene, × 8.

VARIATIONS. Ulbrich (l.c.) has a var. *stolonifera* from Kilimanjaro and Mt. Kenya characterised (*e descr.*) by being caulescent and stoloniferous. Caulescent variants occur occasionally but we have seen no stoloniferous specimens, and no stolons are present in the isotype (*Uhlig* 138 in EA !). There is great variation in leaf size, total leaf-length ranging from 2–40 cm. ; and in the degree of indumentum development. The leaflet cutting ranges in depth and width from almost rounded short lobes and teeth (as in material from Kigezi, Muhavura) to narrowly oblong segments (as in a specimen from Livingstone Mts.). A very dwarf plant, with leaves 2–5 cm. long and pedicels 0.5–3 cm. long has been called var. *gunae* (Schweinf.) Ulbr. l.c. At the other extreme the var. *lanuriensis* De Wild. (Pl. Bequaert. 1 : 39 (1923)) has leaves 19–40 cm. long and pedicels up to 10 cm. long. The var. *genuinus* Ulbr. l.c. is reduced by Robyns (F. P. N. A. 1 : 173 (1948)) to var. *lanuriensis*. With our present know-ledge we doubt if any of the above names can be attached to genetically distinct variants. The specimens so named appear to be habitat forms of a widely ranging species. A possible exception is the material from the Livingstone Mts.

3. **R. stagnalis** [*Hochst. ex*] *A. Rich.*, Tent. Fl. Abyss. 1 : 5 (1847) ; F.C.B. 2 : 177, t. XIV (1951). Type : Abyssinia, Semen, *Schimper* 554 (B, holo. †, K, iso. !)

Perennial herb with suberect or procumbent stems. Elongated cylindrical somewhat thickened roots. Stems 0·6–3·3 dm. long, with few scattered hairs or glabrous, often little branched. Basal leaves with blades ternately palmatisect, segments pinnatilobed or pinnatisect or coarsely toothed, the lobes, segments, or teeth obtuse to rounded, with long spreading hairs on lower surface, glabrous or nearly so on upper surface ; petiole up to 20 cm. long. Sepals spreading, caducous, glabrous. Petals generally 7–8, oblong, narrow, 5–10 mm. long, about 2–3 mm. broad, white. Achenes in spherical heads, on pedicels reflexed in ripe fruit, persistent beak short, bordered margin very narrow and inconspicuous or absent, sides flat or rounded, smooth. Fig. 4/3.

UGANDA. Ruwenzori, Bujuku, Aug. 1931, *Fishlock & Hancock* 69 !
KENYA. Elgon, 12 May 1948, *Mrs. J. Adamson* 484 !
TANGANYIKA. Kilimanjaro, saddle between Kibo and Mawenzi, 17 June 1948, *Hedberg* 1256 !
DISTR. **U2** ; **K3** ; **T2** ; Abyssinia, eastern Belgian Congo
HAB. Bogs and stagnant waters in upland moor, 3000–4750 m.

SYN. *R. vulcanicola* Staner in B. J. B. B. 15 : 314 (1939). Type : Belgian Congo, Kivu District, Kabara, *Louis* 5276 (BR, holo. !)

4. **R. raeae** *Exell* in J.B. 73 : 262 (1935) ; F.C.B. 2 : 177 (1951). Type : Tanganyika, Njombe District, Livingstone Mts., Milo, *M.A.Rae* A 6 (BM, holo. !)

Perennial herb with erect stems, and thickened more or less elongated and tuberous roots. Stems 2–6 dm. tall, loosely pilose, branched. Basal leaves with blades obovate, suborbicular, or oblate in outline, 3–5 palmatisect or palmatilobed, with few, mostly coarse teeth which, with the apices of the segments or lobes, are acute, subacute, or apiculate, broadly cuneate to slightly cordate, up to 5 cm. long and 5·5 cm. broad, adpressed pilose when young, later glabrescent ; petiole up to 13 cm. long. Sepals spreading, pilose on outer surface, caducous. Petals generally 5, yellow, 8–10 mm. long. Achenes in a spherical head, persistent beak straight or slightly curved, margin bordered, sides flat, smooth. Fig. 4/4.

TANGANYIKA. Ufipa District : Ufipa Plateau east of Malonje, 20 Dec. 1934, *Michelmore* 1077 !
DISTR. **T4**, 7, 8 ; Northern Rhodesia, Belgian Congo
HAB. Wet places in upland grassland, 1500–2250 m.

SYN. *R. zernyi* Ulbr. in N. B. G. B. 15 : 714 (1942). Type : Tanganyika, Songea District, Matengo highlands, Lupembe Mts., *Zerny* 166 (W, holo. !)

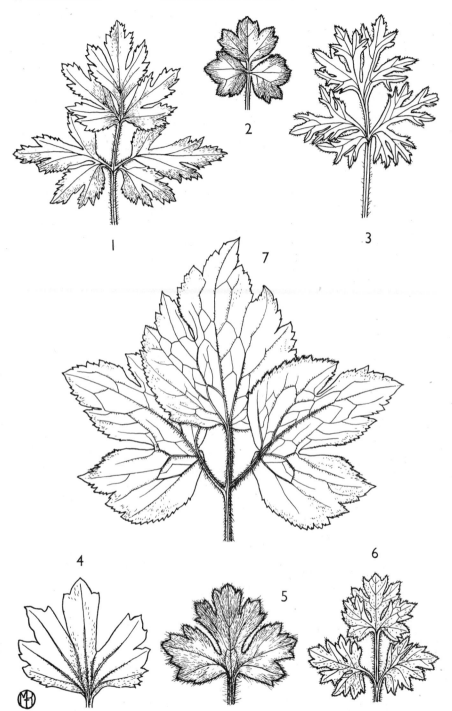

FIG. 4. BASAL LEAVES OF *RANUNCULUS*, all (except 5) × ⅔ − 1, *R. multifidus*; 2, *R. aberdaricus*; 3, *R. stagnalis*; 4, *R. raeae*; 5, *R. cryptanthus* (× 1); 6, *R. keniensis*; 7, *R. bequaertii*.

5. **R. aberdaricus** *Ulbr.* in N.B.G.B. 10 : 904 (1930). Type : Kenya, Aberdare Mts., *Fries* 2566 (UPS, holo. !, S, iso. !)

Perennial herb with prostrate stems, up to 2 dm. long, with solitary flowers at the ends of the branches. Basal leaves trifoliolate, the terminal leaflet with or without a petiolule ; leaflets broadly ovate to oblate in general outline, more or less trilobed and coarsely toothed, terminal 1·4–2 cm. long, 2·2–2·5 cm. broad, lateral leaflets sessile, long adpressed hairs on both surfaces, ciliate at the margins ; petiole up to 1·3 dm. long. Sepals spreading, sparsely hairy on the outer surface, glabrous on the inner surface. Petals probably white, up to 8 mm. long. Fig. 4/2, (p. 17).

KENYA. Aberdare Mts., lower bamboo zone, 31 Mar. 1922, *Fries* 2566 ! ; Sattimma, *Hagenia* zone, 30 Mar. 1922, *Fries* 2897 !
DISTR. K4 ; endemic in Aberdare Mts.
HAB. Open places and by streams in moist bamboo thicket, c. 3000 m.

6. **R. cryptanthus** *Milne-Redhead & Turrill* in K.B. 1951 : 147 (1951). Type : Uganda, Mbale District, Elgon, *Hedberg* 1005 (UPS, holo. !)

Perennial herb, often with a rather stout rootstock, stems up to 9 cm. long but sometimes scarcely any above soil level, with 1 to 3 flowers. Basal leaves trisect, the segments with cuneate, often rather broad, bases ; segments elliptic to oblate in general outline often more or less trilobed and coarsely toothed, terminal 8–21 mm. long, 7–21 mm. broad, dense more or less adpressed hairs on both surfaces, ciliate at margins ; petiole up to 5·5 cm. long. Sepals spreading, with long silky hairs on the outer surface. Petals yellow, up to 4–4·5 mm. long. Achenes in spherical heads on pedicels reflexed in ripe fruit, short beaks, smooth faces. Fig. 4/5, (p. 17).

UGANDA. Elgon, in the crater, 19 May 1948, *Hedberg* 1005 !
KENYA. Elgon, 22 Feb. 1935, *G. Taylor* 3542 !
DISTR. U3 ; K3 ; endemic on Elgon
HAB. Damp places in upland moor, 4050–4100 m.

7. **R. keniensis** *Milne-Redhead & Turrill* in K.B. 1950 : 389 (1951). Type : Kenya, Mt. Kenya, *C. G. Rogers* 615 (K, holo. !, EA, iso.!, BR, iso. !)

Perennial herb, with stems up to 2·4 dm. long, with usually 3 to 4 flowers. Basal leaves trifoliolate, the terminal and lateral leaflets with petiolules up to 3·5 cm. long ; leaflets broadly ovate to oblate in general outline, more or less trilobed and coarsely toothed, terminal 0·9–2·5 cm. long, 1·0–2·9 cm. broad ; with adpressed hairs on both surfaces, ciliate at the margins ; petiole up to 1·4 dm. long. Sepals reflexed, sparsely hairy on the outer surface, glabrous on the inner surface. Petals yellow, up to 7 mm. long. Achenes in spherical heads but sometimes only a few produced from one flower, 4·5 mm. long, with a nearly straight beak. Fig. 4/6, (p. 17).

KENYA. Mt. Kenya, north-west face at base of Cæsar's Seat, 12 June 1933, *C. G. Rogers* 615 !
DISTR. K4 ; endemic on Mt. Kenya
HAB. Peaty or marshy ground in upland moor, 3000–3660 m.

8. **R. bequaertii** *De Wild.*, Pl. Bequaert. 2 : 35 (1923) ; F.C.B. 2 : 178 (1951). Type : Belgian Congo, Ruwenzori, *Bequaert* 3575 (BR, holo. !)

Perennial herb with erect stems, up to 12 dm. tall, somewhat branched into rather loose inflorescences. Basal leaves trifoliolate ; leaflets broadly ovate in general outline, more or less distinctly trilobed, on the lateral leaflets the lobing asymmetric, 7–8 cm. long, 6–6·5 cm. broad, long adpressed hairs on both surfaces, ciliate at the margins ; petiole up to 4 dm. long. Sepals reflexed, sparsely hairy on the outer surface, glabrous on the inner surface. Petals 5, yellow, up to 5 mm. long. Achenes in a spherical head, persistent beak short, hooked, margin bordered, sides smooth, flat. Fig. 4/7, (p. 17).

UGANDA. Ruwenzori, Kasaiga–Bwamba Pass, 11 Jan. 1932, *Hazel* 143 ! ; Kigezi District : Bukimbiri, May 1950, *Purseglove* 3390 !
DISTR. **U2** ; Belgian Congo, endemic on Ruwenzori and in the Virunga Mts.
HAB. Shady places near streams in upland evergreen bushland, 1800–2200 m.

SYN. [*R. extensus* (Hook, f.) Schube sec. Staner in B. J. B. B. 15 : 315 (1939) ; Robyns, F. P. N. A. 1 : 177 (1948), non (Hook. f.) Schube]

9. **R. multifidus** *Forsk.*, Fl. Aegypt. Arab. 102 (1775) ; Robyns, F. P.N.A. 1 : 174, tab. XV (1948) ; F.C.B. 2 : 175 (1951). Type : Arabia, near Taäs, *Forskål* (C, holo. !)

Perennial herb most often with erect stems, 1–12 cm. tall, rarely stems more or less prostrate and rooting at some of the nodes, pilose with adpressed hairs pointing upwards ; flowering stems much branched in the upper part with numerous relatively small typically " buttercup " flowers. Leaves variable, bi- or tri-pinnatisect, the final segments coarsely and irregularly toothed, strongly to weakly hirsute or pilose. Sepals reflexed. Petals 5, yellow, 3–7 mm. long. Achenes in a spherical to slightly elongated head, persistent beak short to well developed and hooked, margin bordered, sides flat, generally with small scattered tubercles but sometimes smooth. Fig. 4/1, (p. 17).

UGANDA. Kigezi District : Behungi, 19 Dec. 1933, *Thomas* 1057 !
KENYA. Mbagathi district, 1 mile S. of Bahati, 14 Nov. 1932, *C. G. Rogers* 41 !
TANGANYIKA. Bukoba District : Karagwe, Nyaishozi, Dec. 1931, *Haarer* 2389 !
DISTR. **U2–4** ; **K3–6, T1–8** ; Arabia to Cape Province, and west to Nigeria and Angola
HAB. Wet ground near rivers, streams and ditches, moist slopes in bushland, grassland or on margins of forest, 1170–3450 m.

SYN. *R. pubescens* Thunb., Prod. Pl. Cap. 94 (1800). Type : Cape Province, *Thunberg* (UPS, holo.)
    R. forskoehlii DC., Syst. 1 : 303 (1817–8), *nom. illegit.* Type : as *R. multifidus* Forsk.
    [*R. pinnatus* Poir. sec. Oliv., F.T.A. 1 : 9 (1868), non Poir.]

VARIATION. Field investigations of populations may enable intraspecific variants to be distinguished within this widely ranging variable species. Attention is particularly called to some low-growing specimens with creeping stems from Uganda (e.g., Ruwenzori, 3210 m., *Purseglove* 297 !)

R. × fratrum *Ulbr.* in N. B. G. B. 10 : 912 (1930) is said to be *R. oreophytus* × *R. multifidus* (*R. pubescens*). It is recorded from about 2500 m. on the western slopes of Mt. Kenya, 29 Jan. 1922, in a damp grassy place by a stream, *Fries* 1247 (K, iso. !) The characters are described as approximately intermediate between those of the putative parents.

7. DELPHINIUM

L., Gen. Pl., ed. 5, 236 (1754)

Annual or perennial herbs with spirally arranged palmatinerved or palmatisect leaves. Flowers irregular. Sepals 5, petaloid and the most prominent feature of the flowers ; the back sepal produced into a well-marked spur. Petals 2–5 (or more), the 2 back with spurs more or less connate and fitting into the calyx spur, the others not spurred and sometimes not longer than the filaments. Stamens indefinite in number, often with flattened filaments. Carpels 1–5, with numerous ovules, sessile, ripening to follicles.

Flowers 4·5–6 cm. diam., sweetly scented :
  Flowers 6 cm. diam., sepals white sometimes with a
    green spot, filaments glabrous or nearly so   .   1. *D. leroyi*
  Flowers 4·5 cm. diam., sepals blue or mauve,
    filaments strongly ciliated   .   .   .   .   2. *D. wellbyi*
Flowers 2·5 to about 3 cm. diam., no odour :
  Flowers about 3 cm. diam., spur rather stout,
    straight and pointing upwards, 2·5–3 cm. long,
    sepals deep blue to green or mixed blue and
    green .   .   .   .   3. *D. macrocentrum*
  Flowers about 2·5 cm. diam., spur stout, spreading
    or pointing downwards, 0·7 cm. long, sepals blue
    or nearly white   .   .   .   .   .   .   4. *D. dasycaulon*

1. **D. leroyi** [*Franch. ex*] *Huth* in E.J. 20 : 474 (1895) ; F.C.B. 2 : 169 (1951). Type : Kilimanjaro, *R. P. Leroy* (P, holo.)

An erect herb 4 to 8 dm. tall, sometimes with the stems having few fine scattered hairs in the lower part, densely softly pubescent in the upper part and on the inflorescence branches, sometimes densely pubescent throughout. Leaves variable in size and degree of segmentation of blades, upper segmented nearly to base. Final inflorescence branches 4–11 cm. long. Flowers about 6 cm. in diameter, with delicate odour. Sepals white sometimes with a green spot ; spur 3–4 cm. long, sometimes S-curved. Filaments glabrous or nearly so. Carpels and follicles 3, with short silky pubescence. Fig. 5.

UGANDA. Mbale District : Bugishu, Walasi, 22 Aug. 1932, *Thomas* 278 !
KENYA. Kilimanjaro, N. of Laitokitok, *C. G. Rogers* 412 !
TANGANYIKA. Rungwe District : W. Mporotos–Rungwe, 13 Aug. 1933, *Greenway* 3552 !
DISTR. **U**3 ; **K**4, 6 ; **T**2, 3, 7, 8 ; southern A.-E. Sudan (Imatong Mts.), northern Nyasaland, Ruanda-Urundi and eastern Belgian Congo
HAB. Upland grassland and evergreen bushland ; 1300–2900 m.

SYN. *D. goetzeanum* Engl. in E. J. 30 : 308 (1901). Type : Tanganyika, Ubena, Ruhudji river, *Goetze* 804 (B, holo. †, K, iso. !)
  D. candidum Hemsl. in Bot. Mag. t. 8170 (1907). Type grown at Kew from seeds collected in Elgon area (K, holo. !)
  D. gommengingeri [Volk. ex] Engl. in E. J. 45 : 266 (1910). Type : Tanganyika, Kilimanjaro, *Gommenginger* (B, holo.)
  D. macrosepalum Engl. l.c. 267 (1910). Type : Tanganyika, Wanegehochland, between Olmoti and Ossirwa, *Jaeger* 431 (B, holo.)
  D. vanderweyeri (no authority) in Gard. Chron. 3rd ser., 54 : 55 (1913). No type.
  Attention is called to the variation in stem indumentum. Field studies and more careful, wide collecting may enable a number of subspecies or varieties to be distinguished on this and other characters.

2. **D. wellbyi** *Hemsl.* in K.B. 1907 : 360 (1907). Type : Abyssinia, between Harrar and Addis Ababa, *Wellby* (K, holo. !)

An erect herb, 6 to 12 dm. tall, glabrous or pilose-pubescent in lower part, more or less densely pubescent in the upper part and on the inflorescence branches. Leaves variable, blades deeply lobed or segmented, sometimes nearly to base, with the segments varying in width. Final inflorescence branches 6–16 cm. long. Flowers 4·5 cm. in diameter, sweetly scented. Sepals blue or mauve ; spur 3·5–4 cm. long, more or less simply curved. Filaments definitely ciliated. Carpels 3, covered with a very dense silky indumentum.

KENYA. Meru District : Lari [Lare], *Ward in C.M.* 3247 ! Cultivated (as an annual), Elgon, near Endebess (*Tweedie* 722 !) and elsewhere.
DISTR. **K**4 ; Abyssinia
HAB. Upland grassland, 1200–1500 m.

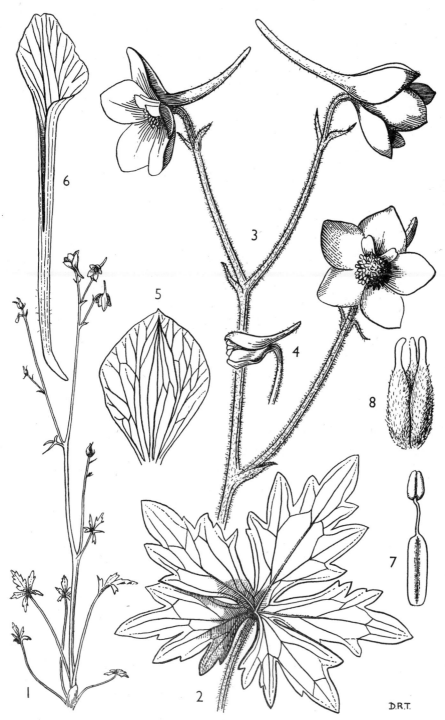

FIG. 5. *DELPHINIUM LEROYI*—1, plant, × ⅛; 2, leaf-blade, × 1; 3, inflorescence, × 1; 4, flower-bud, × 1; 5, front sepal, × 2; 6, back petal, × 2; 7, stamen, × 4; 8, pistil, × 6.

Syn. *D. ruspolianum* Engl. in E. J. 45: 267, fig. 1, F-J (1910). Type : Abyssinia, Galla Highlands, *Riva* 1218 (B, holo.)

3. **D. macrocentrum** *Oliv.* in J.L.S. 21 : 397 (1886) ; Hook., Ic. Pl. tab. 1501 (1886) ; Bot. Mag. tab. 8151 (1907) ; Ulbr. in N.B.G.B. 10 : 899 (1930). Type : Kenya, Laikipia District, Laikipia, *J. Thomson* (K, holo. !)

An erect herb, 9–18 dm. tall, stems glabrous to softly pubescent in lower part, pubescent in upper part and inflorescence branches. Leaves variable but generally with deep and narrow to very narrow and even linear segments. Final inflorescence branches 2–5 cm. long. Flowers about 3 cm. diam. Sepals deep blue to green or mixed blue and green ; spur typically pointing upwards, stout and straight, 2·5–3 cm. long. Carpels and follicles 3, densely pubescent.

Uganda.   Mbale District : Elgon, Mudangi, 12 Nov. 1933, *Tothill* 2411 !
Kenya.   Kericho District : Sotik Hills, 22 Aug. 1946, *Greenway* 7854 !
Distr.   U3 ; K1, 3–6 ; endemic on Elgon and in the Kenya Highlands
Hab.   Upland grasslands and margins of moist bamboo thickets, 1800–3900 m.

4. **D. dasycaulon** *Fresen.* in Mus. Senckenb. 2 : 272 (1837) ; F.T.A. 1 : 11 (1868) ; F.C.B. 2 : 168 (1951). Type : Abyssinia, Semen, *Rüppell* (FR, holo.)

An erect herb up to 1 m. or more tall, stems glabrous to densely pubescent in lower part, densely softly pubescent in upper part and inflorescence branches. Leaves variable in depth and width of segments of blades, generally cut two-thirds or more to base. Final inflorescence branches 0·4–1·5 cm. long. Flowers about 2·5 cm. diam. Sepals blue, violet, pinkish or almost white ; spur stout, 0·7 cm. long, nearly or quite straight. Carpels and follicles 3, shortly and densely pubescent.

Tanganyika.   Ufipa District : Ufipa Plateau, River Nsanvia, south-west of Mwazye, 21 Apr. 1934, *Michelmore* 1041 !
Distr.   T4, 7 ; A.-E. Sudan, Eritrea, Abyssinia, Northern Rhodesia, Nyasaland, Belgian Congo, Cameroons
Hab.   Upland grassland and *Brachystegia* woodland, 1440–1800 m.

It is strange that no material of the species has been received from Kenya or Uganda.

INDEX TO RANUNCULACEAE